The
BARN BOOK

Title page: *These carefully braced double doors have been aged by countless seasons.* **This page, above:** *An animal-size door, a people-size door, and a hinged, drop-down haymow door serve different functions in a working barn.* **Right:** *The silver patina of the siding and peaked cupola lend an elegant bearing to this utilitarian building.* **Below:** *Weathered barn wood comes in an artist's palette of subtle colors, accented by the patterns of its grain and highlighted by knots and whorls.*

Roofing a barn has never been a job for the faint of heart. **Below:** Preconstructed bents are moved into place and then raised by pulleys, shoulder power, and the push of pikes. **Following pages:** Wooden pegs called treenails are hammered in to connect each mortise and tenon. A red barn with white trim is as American as apple pie.

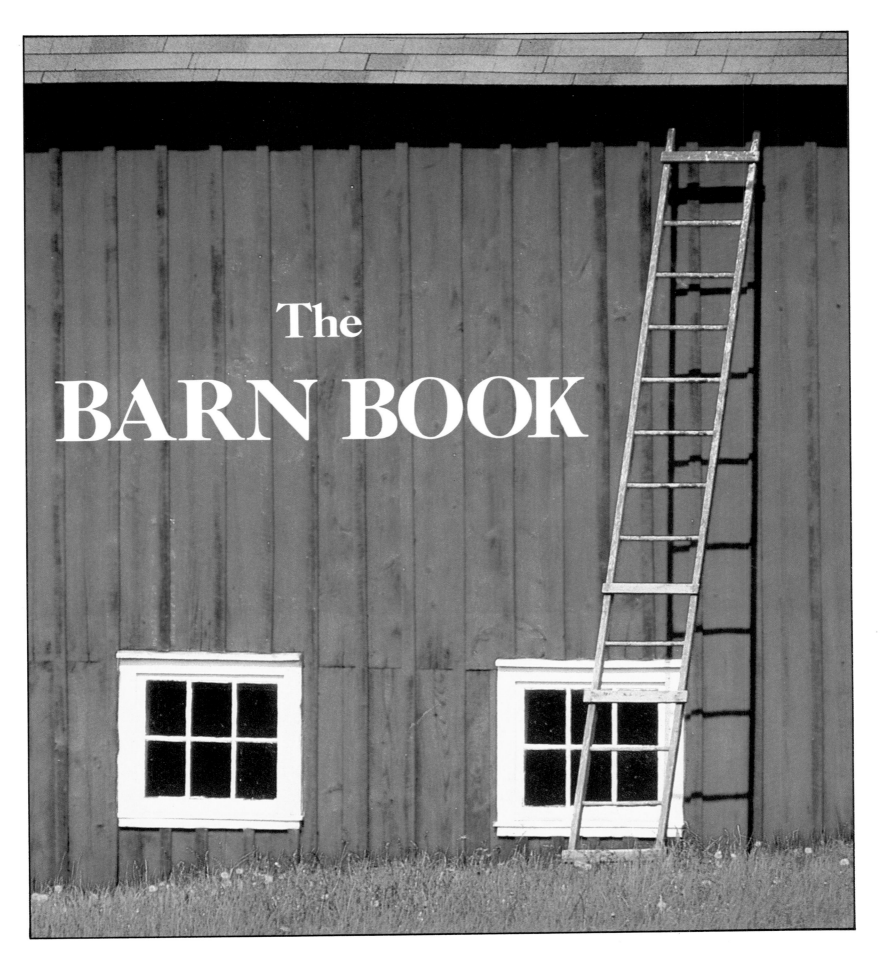

The
BARN BOOK

GALLERY BOOKS
An Imprint of W. H. Smith Publishers Inc.
112 Madison Avenue
New York City 10016

This edition first published in U.S.
in 1990 by Gallery Books,
an imprint of W.H. Smith Publishers, Inc.
112 Madison Avenue, New York, New York 10016

ISBN 0-8317-0687-2

Printed and bound in Spain

For rights information about the photographs in
this book please contact:

The Image Bank
111 Fifth Avenue, New York, NY 10003

Producer: Solomon M. Skolnick
Author: Carolyn Janik
Designer: Ann-Louise Lipman
Editor: Joan E. Ratajack
Production: Valerie Zars
Photo Researcher: Edward Douglas
Assistant Photo Researcher: Robert V. Hale
Editorial Assistant: Carol Raguso

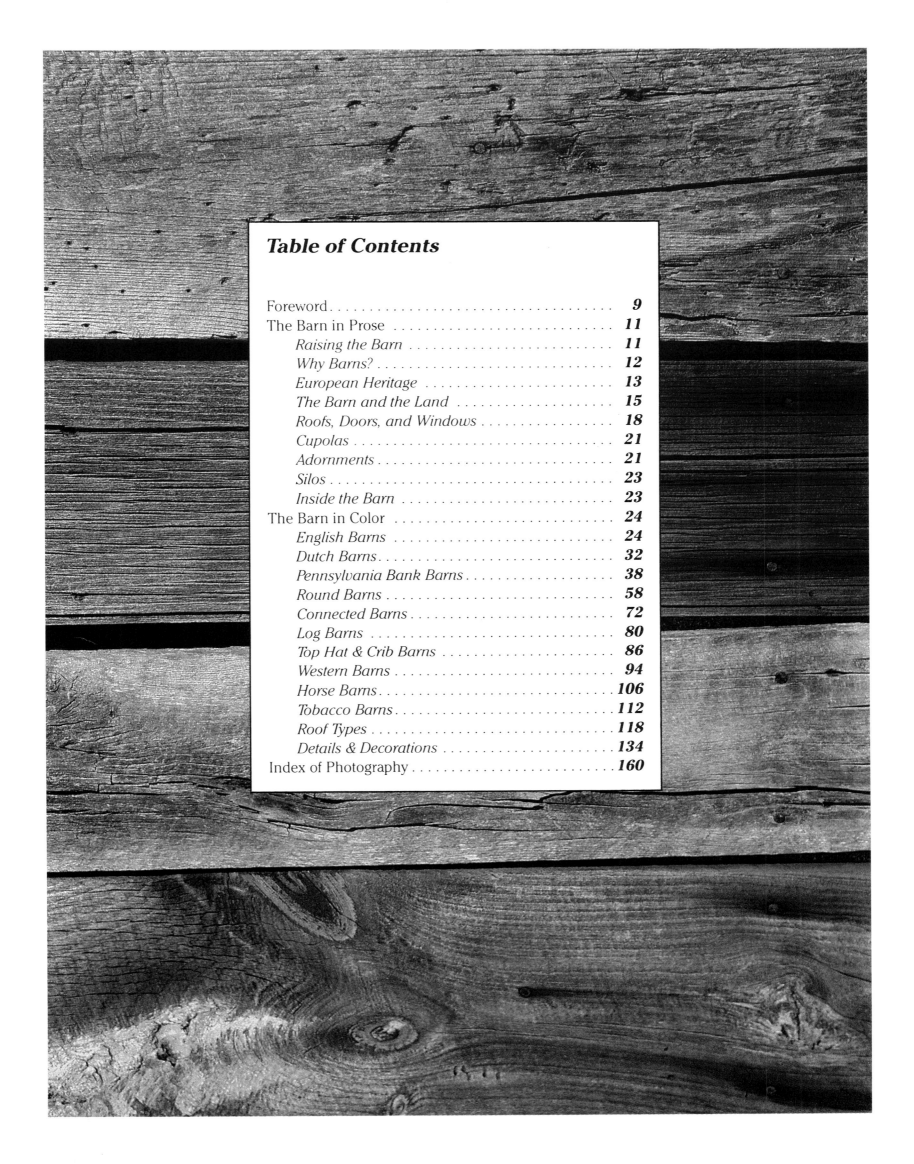

Table of Contents

FOREWORD

When the subject is barns, almost everyone has a story to tell. It may be recollections of chilly mornings milking cows, or hot summer afternoons stacking scratchy hay. It may be games of hide and seek, or flying through a cavernous haymow on a rope swing. Some remember barn dances, when neighbors would gather on a summer evening to visit and dance in the empty haymow before it was once again stacked with hay for the winter. Just the smell of a barn is enough to bring back a flood of memories.

Even city dwellers who have never set foot in a barn have their stories to tell. Almost everyone has seen a barn, and once seen, these magnificent structures are hard to forget.

To the majority of Americans today, barns are a feature of the landscape, architecture admired on a Sunday drive through the country. They are handsome, picturesque relics, reminders of days past, subjects for photographs, paintings, even poems.

To the farmers who own and use them, barns have an entirely different meaning. The barn is a place of work, the place where livestock is kept warm and dry, hay and grain are stored, and machinery is repaired. Barns are working buildings, built to serve a practical purpose on the farm.

We treasure barns for their simple, functional design, for the beauty of their lines, for their sheer size, for their massive frames and cathedral-like interiors, for their weathered wood and carefully fitted stone. We treasure them as symbols of a rural past from which most of us are now far removed. We treasure them because of what they can teach us about the history of a farm and what farming was like in days past. We treasure them as the center of life and activity on our farms today.

American barns of 200 years ago were simple wooden structures used to store grain or shelter animals; today's barns are low-lying pole buildings, sheathed in metal, and equipped with a battery of high-tech equipment. In between, farmers built barns of all shapes and sizes: round, polygonal, square, rectangular; from one to four stories tall; of log, milled lumber, stone, brick, and even adobe. Some are very plain, with nothing but the essential four walls and roof. But architectural styles and ornamentation sometimes made their way to barns. Barns can be found with Italianate scrollwork, Gothic pointed gables, and even wooden false-fronts which make the building look much bigger and more impressive than it really is.

Most barns are anonymous buildings. Few were designed by architects. Many barn designs were influenced by the ethnic traditions of the farmers who built them, and the materials available locally. Frequently, professional barn builders developed a distinctive style which can be recognized throughout a certain locality. In the early 1900's, barn plans designed by agricultural engineers at land grant universities and distributed through the cooperative extension service, increased standardization in barn design. Barns were also available ready-made through mail order suppliers such as Sears & Roebuck.

Along with stories about barns, almost everyone laments the disappearance of traditional barns from the American landscape. We hate to lose even one. After all, they are a limited resource, treasures that cannot be replaced. If we lose these treasures, the image of the American farm will never be the same.

With the decline in the number of individual farms and in the total acreage of farmland in the United States over the past several decades, hundreds of farmsteads have been abandoned or demolished. In the countryside across North America, barns are bulldozed to make way for new shopping centers and housing developments. Some may be saved and adapted for homes, stores, or community centers, but only a few barns will survive without farms.

Even on working farms and ranches, many historic barns face an uncertain future. Most of the barns pictured in this book were constructed to accommodate agricultural practices that have long ago been abandoned, such as the hand threshing of grain or the hand milking of cows. When the tractor replaced the draft horse, farmers no longer needed stables for their horses or places to store their horses' feed. As farms became more specialized, many farmers sold off their livestock. The barn that was designed to shelter animals and to store huge quantities of hay sat empty.

Fluctuations in the farm economy and changes in particular farming industries have also taken their toll on historic farm buildings. Unused dairy barns in New England and empty tobacco barns in the Southeast are relics of once thriving industries.

As changes in agricultural practices made older barns obsolete, new, metal-clad pole buildings began to take their places. Many barns were relegated to the status of an attic—places to store old tires, odds and ends of lumber, and everything else the farmer didn't know what to do with. The chore of painting the old barn and patching the roof seemed too time-consuming and expensive for a building that wasn't used very much. Before they knew it, the barn was gone—blown down in a tornado, collapsed under the weight of snow one winter, burned by the fire department for practice.

In spite of the losses, the outlook for barns is not all bad. Many farmers have found that these big buildings are very versatile. As farming has changed through the years, some farmers have adapted their older barns to fit their current needs: threshing floors can be used to store hay and equipment, stalls can be rearranged for different types of livestock, and openings can be enlarged or moved to accommodate new machinery. Even state-of-the-art mechanical systems can be installed in 150-year-old barns. In fact, if a farmer needs a building for any type of operation, chances are the old barn can be adapted to fill the need.

Economic crises are often boons for preservation, and the farm crisis of the 1980's was no exception. Facing decreasing profits and high interest rates, many farmers could no longer afford the new facilities that proliferated during the 1970's. Instead they chose to conserve resources by adapting existing buildings to meet their needs.

On farms all across the country, dairy barns are housing hogs and sheep, haymows are storing grain, combines are parked where draft horses once stood. A 100-year-old barn in Illinois houses a grade-A dairy. A historic three-story round barn in Ohio is used to clean, sort, and package soybean seeds. A 1900 hay and livestock barn in Missouri is now a roomy farm shop.

Barns are being converted to less traditional farming uses as well. In California, several old dairy barns are now wineries. In Kansas, an old barn serves as a farmers' market. In Iowa, a farmer uses his 1933 clay-block barn for Christmas tree sales.

A national barn preservation program known as BARN AGAIN! has helped thousands of farmers all across the country put their barns back to work. An effort of the National Trust for Historic Preservation and *Successful Farming* magazine, the BARN AGAIN! program has helped to develop new uses for outdated barns, and has shown that using an old barn can be cost effective.

Anyone who enjoys barns, anyone with a barn story to tell, will enjoy taking a trip through rural America by traveling through the pages of this book. You will witness the incredible range of barn types and styles found throughout the country. And hopefully you will be inspired to join the effort to preserve our great American barns.

RAISING THE BARN

The Schneiders were building a new barn and had been preparing the foundation and the superstructure for many weeks. Virtually all the members of their surrounding community were coming today to help them "raise the barn." As the first wagonload of neighbors came rumbling down the road to the farm, most work stopped as the Schneider family went to greet their neighbors.

But one man, Bob Clayton, the master carpenter the Schneiders had hired two months ago, never even broke the rhythm of the 40 pound mallet he was swinging. Once he got this tenon into its mortise on the sixth bent, or frame, he would fasten it with the oak pegs he carried in his pockets. Then he would move right on to assemble one more of the massive post-and-beam structures. Within the hour, he would have them joined together to create the seventh and last bent.

Every family in the community was coming to the Schneider farm to help in lifting, moving, positioning, and bracing the H-shaped skeletal sections of the barn that Bob Clayton and the Schneider men had hewn from logs and fastened together.

Work on the barn had actually started in March, almost seven months ago, when Alphons Schneider had chosen and marked out the site. He had picked a knoll not far from the house but also close to the stream. The slightly higher ground would facilitate drainage away from the barn and the brook would be an excellent source of water for the oxen.

Throughout the spring and summer, the entire family had worked at digging the hole for the foundation during every moment that could be snatched from other chores. Large stones, come upon while clearing the land and preparing the fields, had been carefully assembled and stacked to create a foundation. More stone had been cut, chipped, and fit together to build the three pillars that were carefully centered down the length of the foundation excavation. These columns would support the almost 40-foot-long center floor beam that was the main girder, the very spine of the barn.

The older Schneider children, Eric, John, and Mary, had each measured and marked the two places in the foundation that would be niched out to hold the ends of this beam, and Bob Clayton had checked their measurements again. There could be no mistakes. The great oak center beam, almost 12 inches square in section, had to be secure in the foundation and rest squarely on its supports, for it would carry the weight of the joists and the floorboards as well as much of the superstructure of the barn. The center beam would also carry the weight of wagons loaded with hay and grain, of cattle and oxen, and of many people doing their chores, threshing, spinning, sometimes even dancing.

Once the foundation was completed and the center beam was set in place, the Schneiders and two hired men laid the beams of the sill, joining them at the four corners of the barn. Each of these massive beams (only slightly smaller than the main girder) had been cut from a different tree and hewn square with an adz. Now the four together made the perimeter of the barn, the sill to which the posts of the bents would be fastened.

Bob Clayton had carved notches in the two sill beams that ran the length of the barn. The notches would secure and support the floor joists, which would be braced at their midpoints by the huge center girder on its pillars of stone. Floorboards to rest on these joists had been cut and shaped and then left to season, waiting for raising day when they would be laid to provide a working surface.

While the foundation of the barn was being built, Bob Clayton had been busy using his training and talent to create the seven bents that would be the vertical framework of the barn. Each bent had to be measured and cut so that it would come out to be exactly the width of the barn when beams and posts were securely interlocked. This was the job he was finishing now on the last bent.

To create the barn's infrastructure, the ends of all of the horizontal beams had to be carved into tenons. This was done by cutting away the wood until a rectangular projection was created. Vertical posts had mortises, or holes, carved into them that would receive the tenons. Then the tenons on the beams and the mortises on the posts were joined with the help of 40 pound mallets called beetles. Treenails, sometimes called trunnels (round pegs, usually of oak), would be driven through the aligned holes in the mortise area and its interlocking tenon to permanently fasten the beams and posts together. When all the beams and posts of the Schneider's barn were pounded together, seven great H-shaped bents were lying on the ground near the foundation.

The men who came to the raising would spend the better part of the day moving the seven bents into place, lifting each one to a vertical position, and then bracing them to the sill. Once the bents were standing and secure, forty-foot beams, called plates, would be placed on top of the end posts. These beams would run the length of the barn and unite all the bents. At each intersection of plate beam and bent post, diagonal braces would be added to ensure stability.

After all the preparations were completed, the first gang of men lined up and lifted the bent as a team, their shoulders to the wood. When shoulders and arms could reach no further, a second gang of men aided them with pikes, 16- to 18-foot-long wooden poles with a metal spike at the end. This second gang stood behind the first and pushed the bent up higher with the poles. Meanwhile, men on the opposite side of the bent helped raise it by pulling on ropes rigged to a pulley system.

Once the bent was upright, it was quickly supported by men and boys on both sides of it while still other men rushed in to brace it to the sill. As each additional bent was brought up to its vertical position, supporting beams, called struts or girts, were set between the posts at about halfway up the height, tenon into mortise. When all the bents, braces, and girts were in place, the plate beams were set atop the bents.

Now the most agile (and perhaps the bravest) of the gathered men climbed up to the anchor beams on the newly erected skeleton and set the braces that reinforced the joinings of the plate beams and the posts. Finally, the rafters of the roof were set into the plates and joined at the roof ridge.

The work, with all its pulling, pushing, hammering, sweating, shouting, and swearing, was then over. In the afternoon shadows of the standing timbers that would soon be dressed out into a barn, it was time for the great eating and drinking to begin.

The women, who during the raising process had been at work preparing a great feast and setting up the tables and log benches, sat down with the men and children. Someone led the group in saying grace, Al Schneider stood and thanked them all, and the community began eating and talking and later, dancing, to celebrate a job well done. The celebration continued until well after the sun had set.

The next morning, Mary Schneider sat at the back corner of the foundation with a hammer and chisel in her hands. She was etching yesterday's date, September 17, 1841, into the largest stone in the foundation. Her father, her older brothers, Bob Clayton, and the two hired men were already nailing into place the vertical boards that would be the outside sheathing of the barn.

But why did everyone leave when the barn was only a skeleton? Why didn't they *finish* the work and *then* celebrate?

Because that's the way it was done. Contrary to folk fantasy, American barn raisings didn't start at the break of dawn with a open patch of earth and come to an end at sunset with a completed barn. They were great cooperative undertakings and wonderful parties, but there was much work done both before and after the designated day. The farmer, his sons, and his hired hands did all of the preparation work and most of the finish work, including the outside siding, the doors, and the roofing. Usually the hired master-carpenter stayed on the job and worked alongside the other men for a few days after the raising, since he was rarely paid until the barn was straight and true. But he almost never stayed long enough to see the first load of hay pitched into the mow.

It was also a common practice to chisel barn raising dates into foundation stones or to carve them into supporting timbers. Many barns even have their dates carved in the lintel over the main entryway. On other barns, dates are carved or painted just under the peak of the gable. This careful dating says something about the pride of the farm family in having created the most important building on the farm. It could be said that the date celebrates the birth of the farm as a surviving entity.

WHY BARNS?

About 80 percent of the population of North America now lives in cities or suburbs. If you were to ask any number of these people why farmers have barns, you'd very often hear something like "to shelter their animals." Most barns today do shelter animals, but shelter for animals was not their most important function during the agricultural centuries of our history, nor is it their only function now.

Our word *barn* comes from the Anglo-Saxon *bara-ern,* meaning "barley place," literally a place to keep barley, an important crop in ancient England. The word passes through Old English as *berern* and then Middle English as *bern* before becoming the modern *barn.* But throughout its more than two thousand years in the language, it has always meant a place to store grain. *Webster's New Collegiate Dictionary* defines *barn* as "covered building for storing grain, hay, etc., and also, in the United States, for stables, etc."

"Stables? That's horses. What about the cows?" you might ask. In our colloquial usage, cows live in the barn too, but the correct word for a building housing cows or other cattle is *byre,* an Anglo-Saxon word still in common use in Great Britain. In egalitarian North America, however, barns can house cattle, horses, pigs, sheep, chickens, and stray cats as well as store grain and shelter farm machinery. They are still the heart of the farm, even farms run with the brain power of sophisticated computers.

For early North American farmers, the barn was the most essential building in their lives. There are many stories of pioneer families building the barn before building an adequate house. Particularly in the northern areas of the Temperate Zone (virtually all of Canada and more than half of the United States), barns were absolutely essential for the storage of grain and the survival of the farm animals.

Until near the end of the nineteenth century, there were few feed stores and even fewer grocery stores convenient to the farmstead, and very few mail-order catalogs. The farmer grew and stored what his family and animals would eat throughout the winter, and the barn was his storage place. At harvest time, it was packed to bulging, packed enough, everyone hoped, that the food supply lasted until the new crops were harvested.

The barn was the building that allowed the farm to function, and it was designed and built with function in mind. Nothing else in North American architecture is so purely utilitarian, and in its purity, so simply beautiful. The barns of the United States and Canada are the crowning glory of our vernacular architecture, an architecture created without architects.

EUROPEAN HERITAGE

Each immigrant group brought to the shores of North America its own idea of how a barn should look. When they got here, however, they usually met longer, colder winters and thicker, more abundant forests than they knew in their home countries. Sometimes they found more rocks in the land than soil. So European ideas of barn building were adapted to the new topography, the new climate, and to the needs of the individual farm and farm family. But the basic concept of the barn remained in the minds of each immigrant group, and each created a distinctive architectural style that is easily recognizable in the old barns still standing today and in the new barns that are the great-grandchildren of the early constructions.

The ancient Saxon barns of England, Holland, and Germany were shaped like the inverted hull of a ship. Hay wagons, or wains, entered at the end, since a large enough door could not be cut into the low sides of the barn. The structures were therefore divided longitudinally, with a wide center aisle marked by columns or supports and flanked by narrower aisles on each side. The center aisle was used as a threshing floor. The side aisles were used for hay storage and for animal stalls.

Many architectural historians refer to this design as a basilican plan (after the early Christian churches) and go on to compare the basilican plan barn interior to Gothic cathedrals. They liken its threshing floor to the church nave, divided from the side aisles by pillars at regular intervals just as the posts of the bents that separate the bays of a barn divide the threshing floor from the darker, narrower animal stalls and storage areas. Even the high, steep barn roof brings to mind the soaring cathedral ceiling.

The popularity of the basilican plan in barn construction waned in both England and Germany before the settlement of North America, but it continued much longer in Holland. It was the Dutch, therefore, who brought this design to the New World. Many of the oldest standing barns in North America are the basilican plan Dutch barns of New York and New Jersey.

The New World Dutch barn is always a simple rectangle, almost square, in fact, but wider than it is long to accommodate the wide central aisle with its important threshing floor and the two narrower aisles on either side. Hand-hewn posts separate the center aisle from the sides, and great transverse anchor beams stretch from post to post overhead. Virtually all the early New World Dutch barns were constructed entirely of wood, even the door hinges and roofs. A row or two of stacked stones was used in place of a foundation to keep the sills off the ground and thus prevent rotting.

The roof of this type of barn is always two steep symmetrical slopes, forming gables at either end of the barn. A door, large enough to accommodate a wagon with a full load of hay, is always located in at least one of the ends. In fact, gable-end doors are the distinguishing feature of Dutch heritage construction, and there is usually one at each end of the barn. The second door was added opposite the first to allow the unloaded wagon to drive through the barn and exit without turning around. Windows are few or nonexistent, but ventilation and shafts of light are often provided by pigeon holes at the peak of the gables.

The small English barn actually became popular in

the seventeenth and eighteenth centuries in many parts of Europe, but the English take credit for introducing it in New England, and the name has stuck. It is also known in North America as the Yankee barn, but it can be found not only throughout New England but also in the Midwest, Canada, the Carolinas, and some areas of the Appalachian Mountains. It is also the grandsire of many creative new barn variations throughout North America.

If you are driving North America's back roads and you keep your eyes open for wooden structures that are weathered silver with time, you just might spot an English barn, or a derivative of one. Many English barns were built entirely of wood, with vertical board siding and a gable roof. Others, however, are stone, or brick, or horizontal logs (chinked or unchinked), or board and batten, or clapboard, or rubble masonry. What makes them all English barns is their size (three bays) with the main wagon door in the center. That is, the door is in the center of the long wall of the barn, not the gable end.

The word *bay* might be confusing here, calling up visions of housing-tract windows or familiar bodies of water, but in architecture the word denotes the space between beams, buttresses, or pillars. A three-bay barn has four post-and-beam bents, one at each end and two more dividing the length of the barn into thirds. In the English barn, the threshing floor runs the width of the structure between the two internal bents, as opposed to running the length of the barn, as in the Dutch style. In other words, the center bay was both the threshing floor and the wagon road.

The great wagon doors open to this center bay, and in most North American barns, a wagon could drive right through and exit by another set of doors on the opposite side of the barn. Besides adding to the convenience of the wagon driver, the double sets of doors contributed to the threshing process. When both sets were opened, the wind was used to help in the winnowing of wheat from chaff.

The English barn was often the barn of choice for the small farmer. It is a one-story structure and therefore was easier to build than some of its gigantic cousins, yet it still allowed for the separation of livestock on one side of the threshing floor and grain and hay storage on the other. As farms prospered and more space was needed, hay was sometimes stored in the area under the roof. This loft was created by making a temporary floor of poles suspended across the anchor beams. When still more space was needed, lean-tos were added to the sides of the barn or ells were built protruding from one of its walls.

Some prosperous farmers, not wishing to add on to their existing structures, simply built barnyards of three or four small English barns forming a U or sometimes a courtyard. Each of these barns would then have a specialized function.

Do you remember the farm animal storybooks you treasured as a child? Or the books you've recently read to your children or grandchildren? Think about the barns pictured in them. Big, red buildings on stone foundations. That's the American fantasy barn, and its prototype still stands in southeastern Pennsylvania.

Farmers from the mountainous regions of Bavaria and Switzerland brought with them the concept of using the slope of the land. They found the terrain in their new country different from that of their homeland, but they adapted remembered techniques to their new location, striving always for efficiency and economy. From their ideas and efforts, came the Pennsylvania bank barn (sometimes called the Sweitzer-type barn), the first of the multi-story barns in North America.

This was a barn where animals were warmer in winter and cooler in summer. It was a barn where the entire inside dimension could be counted on as floor area for threshing and grain storage, an area completely separated from the animal stalls, yet with easy access for the wagons. These Pennsylvania barns were the first truly *big* American barns, forerunners of the giants that would stand proudly across the continent as the central and western farmlands were cultivated.

The north side of the Pennsylvania style barn was almost always set against a bank or hill, or the land was excavated out to create a bank. The first story was built of stone, usually the readily available and easy-to-cut local limestone, and set into the bank. The floor of the second story was at grade with the hill, so that a wagon could be driven up the path and enter into the haymow and threshing floor. It was constructed of wood and cantilevered out over the wall below on the south side. The farm animals sheltered in the lower level had easy access to the barnyard through several doorways set in the stone. When working with his animals, the farmer also had some shelter from inclement weather under the overhang of the haymow.

As the original Sweitzer barn plan became more and more popular, it was refined and altered. Many farmers used abundant local stone to build both gable ends, creating imposing and enduring monuments. Other farmers went even farther and built the entire barn of stone, without the overhang.

The advantages of the Pennsylvania bank barn are obvious. With the north side of the lower level built against the earth, the animals were sheltered from the winds of winter and the baking sun of summer. The overhang over the south side also sheltered the openings to the lower level. The haymow, threshing floor, and grain storage areas were huge, easily accessible, and completely separate from the animal housing. The barn's popularity lasted into the beginning of the twentieth century.

THE BARN AND THE LAND

No other structures in North America were so influenced by the climate, topography, available resources, and intended function as our barns. The result is a vast diversity of barn architecture which demonstrates the beauty of functional simplicity and reflects pride in craftsmanship.

Snow is a very real and plentiful part of winter in Maine, New Hampshire, Vermont, and northern Massachusetts, and early farmers didn't have the luxury of sitting by their fireplaces and watching it pile up against the windowpanes. The land could rest, but the animals had to be tended. With typical New England ingenuity, some of these farmers devised a means of getting each day's chores done without ever having to trudge through the snowbanks. They built their farm buildings connected one to the other.

Each building, however, remained distinct, with its own shape and roofline. There was never a question of the farmhouse being a part of the barn. Instead, the buildings meandered away from the farmhouse in an uneven line, turning sometimes to form an L or perhaps an E, or just an interesting zigzag on a hilltop. This was vernacular architecture that bespoke need and function. It was also aesthetically pleasing in its variety of shapes and in the uneven and unexpected rhythm of its lines.

Most concentrated farmsteads can be found today in southern Maine, throughout New Hampshire, and in eastern Vermont, but there are also scattered examples in Massachusetts, Quebec, Ontario, Nova Scotia, and New Brunswick.

Harsh winters were also a problem in Quebec, but Canadian farmers came up with a different solution, one quite obviously drawn from their French heritage. The traditional, small, Breton farmstead sheltered the family's living quarters, the barn for grain storage, the byre, the stable, and the other animal pens under one unbroken roof. The size and shape of the doorways to each of these "apartments" were made quite different, however, and

clearly indicated what one could expect to find inside. From this peasant homestead was born the long, low Quebec connected barn. Most Canadian families, however, lived in a house separate from the multi-purpose barn building.

The rooflines of Canadian connected barns are long and unbroken. Walls divide inside areas, however, and the exterior walls are broken with doors of different sizes at unexpected intervals and windows in no apparent order, except perhaps as an expression of function from the inside. Dormers set into the roof above the haymow doors are quite common, but some Quebec connected barns also sport inexplicable dormer windows.

Until recently, thatched roofs were still common on Quebec barns and exterior walls of log or planks (either vertical or horizontal) were prevalent. Occasionally you might find shingles used on at least some parts of these barns, but the stone construction of their Breton heritage seems to have disappeared.

Even in America's earliest farming days, tobacco was a very profitable crop, and it was grown from the Connecticut/Massachusetts line south to Georgia. Essential to tobacco farming were barns with excellent ventilation for drying the tobacco leaves. Tobacco barns therefore were among the first truly specialized barns in North America. Depending upon the climate and the kind of tobacco being grown, their size, construction, and ventilation varied tremendously from state to state.

The Connecticut River valley seemed to have the perfect soil for the broadleaf tobacco that was used in wrapping quality cigars, and the industry became a roadside landmark of the central area of the state. Because the tobacco leaves have to be grown in the shade, gauzy sheets were stretched above the fields and supported by poles dug into the earth at frequent intervals. The white fabric fluttered in the wind over acre after acre with huge, silver gray, weathered tobacco barns standing like sentinels in snow.

The typical size of a Connecticut River valley tobacco barn was 30 by one hundred feet, large enough to dry the tobacco grown on about three acres of land. Inside were four tiers of braces. Across these braces (connecting two or three) were laid the poles from which the tobacco would hang. Leaves hanging from the lowest of these poles still left a clearance of seven feet or more, so that the wagons loaded with freshly picked leaves could be driven right into the barn.

The tobacco barns of the Connecticut River valley are unadorned, rectangular structures, windowless and unpainted, with simple gable roofs. The vertical boards used as siding, however, were attached in such a way that every other board could be bent outward at the base in order to provide the ventilation needed to properly cure the hanging leaves. This process was an entirely natural one and the rows of tobacco barns along the highway must have been the quietest industrial complex in the world.

As the twentieth century comes to a close, however, the growing and drying of broadleaf tobacco in Connecticut is an almost extinct industry. The chief contributor to its demise was the development of the homogenized tobacco cigar wrapper created from lower-grade pulverized tobacco that has been rolled into sheets not unlike paper. This product works well with automated cigar manufacturing equipment and thus makes cigars less expensive to produce than those that are hand wrapped with the shade-grown and barn-dried leaves. As a result, the tobacco barns of the Connecticut River valley are rapidly disappearing. Some have been abandoned and are looted for their decorative weathered siding, while others are demolished by developers' heavy equipment.

In Pennsylvania, there were far fewer tobacco barns built than in Connecticut, and they were not specialized for drying broadleaf cigar wrappers. These barns are closer in size and shape to the English barns of New England, except that they have roof ventilation systems. They can be identified by what seems to be an extra roof just a few feet long on each side of the ridge line. Under this cap are the roof vents of the barn. These barns also have panels on all four walls that are hinged and can be opened like shutters.

In Maryland, tobacco barns were built high rather than long and look like English barns doubled in height. Many also had lean-to-like extensions added to one or both gable ends. Rather than a roof ventilation system, these barns have panels that open on hinges. Each panel runs from the ground to just under the eaves. When all the panels are opened, the barns seen from a distance seem to be striped.

Some larger Kentucky tobacco barns were actually painted in stripes by using a contrasting color on the hinged ventilation panels. Most of these barns were dark colored, with the panels in a lighter color.

Like the Connecticut River valley tobacco industry, privately owned tobacco barns are also rapidly vanishing from the American landscape. While exploring the back roads of Kentucky, Tennessee, Virginia, the Carolinas, or Georgia, however, you might still come upon some log tobacco barns. In these simple buildings, the logs were left unchinked as a means of ventilation. In Virginia, you might also find English barn style buildings with huge vertical panels of louvers set in the walls on all four sides. These are probably the tobacco barns of small, independent farmers.

The early settlers of the southeastern coast brought with them the concept of the English barn, but they colonized a land where the climate did not demand winter shelter for domestic animals. The primary requirement of their farm outbuildings was the storage of grain. As a result, the southern crib barn came into being.

The wagon doors in the northeastern barns became an open passageway between two separate storage cribs. Above the two cribs and connecting them was a larger, cantilevered haymow. The cribs were usually constructed of logs, while the haymow was sided with vertical planks. Only very rarely was the structure painted.

This simple storage barn worked so well that westward-moving pioneers carried its plan with them into the Appalachian Mountains and beyond.

As farming moved westward, the crib barn took on a shape like no other barn in the world. The English barn concept of three bays survived in the form of a central drive-through bay and an additional bay on each side of it for the storage of grain. Each of these storage cribs is completely enclosed with its own slanted roof. The two roofs could not have met to form a gable over the central aisle because the opening would have been too low for a wagon loaded full of hay to pass under. So the top-hat addition (also called a "top-knot") appeared and became characteristic of these barns. Like a miniature barn above the central wagon passage, this storage space has its own gable roof. When one is looking at the entire roofline of the barn, it seems as though the gable angle of the two cribs is arbitrarily clipped and then raised up to create the roof of the top-knot.

Most of these top-knot crib barns can be found in the south central states from Tennessee to Texas, where grain and hay storage are far more important than animal shelter. They are especially popular in Arkansas, where they are used for drying and storing rice.

Crib barns used for corn and other grains usually have no windows, although some have a haymow door above the wagon entranceway. In some instances, the angled rooflines of the two ground-level storage cribs are extended out even further to form shelter for farm machinery or for animals during stormy weather.

Farming came to the Midwest at least a century after

it was well established along the eastern coast, and the barns of this area reflect the assimilation of the building styles brought from Europe and adapted in America. Since wood was the most plentiful raw material, it is almost certain that the early barns of the American Midwest were made of logs, probably left in the round and chinked with mud, hay, or rags.

As farming became more specialized in this area, farms (and their barns) became larger. Dairy farming in particular presented itself as a viable livelihood, and specialized dairy barns were built. In these barns, the threshing floor disappeared entirely and the haymow got bigger and bigger. Silos also appeared on the landscape. Sometimes shedlike roofs were added that ran the length of the barn and provided shelter for the farmer as he worked with his stock. These extra roofs were especially important because they carried rainwater and the runoff of melting snow away from the foundation of the barn. Large, pointed extensions were also added to the gable ends of the roofs to shelter the haymow doors.

The barns of the north central area of the United States are the giants of the nation. They are younger than eastern barns but as purely functional to the age in which they were created as their older cousins. And some would say that there is as much beauty in the efficiency of the twentieth century as in the craftsmanship of the eighteenth and nineteenth.

By the time farmers staked out their land on the West Coast of North America, the industrial revolution had permanently altered barn construction practices and the railroad could bring otherwise unavailable materials, tools, and supplies to the newly established homesteads. But the settlers came from middle and eastern regions, and they brought with them the concepts of barn that were familiar. In their new homes, they changed what they knew to fit the new land and the new ways of doing things.

As a result, the barns of the West Coast are recognizable offspring of their eastern predecessors, yet they have a style that is endemic to their new region. Settlers found the climate here milder, the crops different, and the way of life generally more relaxed. Existence was less of a struggle, and the western farmer built barns that seem to have a greater serenity of line.

In California in particular, the crib barns of the south central states became larger and took on much longer, gently curved roofs. They often seem to blend into the land, raising gracefully from it to a peak and then descending just as gracefully to meet the land again on the other side of the structure. These western barns seem to stand in quiet harmony almost like miniature mountains in their fields, whereas their eastern cousins stand proudly and staunchly as monumental symbols of man on the land.

No one can be quite certain what made some farmers abandon the age-old rectangular barn shape for the circle. Perhaps on a winter night a farmer calculated that there would be more usable interior space relative to the amount of outside wall that needed to be built. Or perhaps there was a philosophical quest to incorporate the perfect, unbroken shape of the circle into the daily life of the community. Or maybe some people just got to feeling bored and began looking to create something newer, bigger, and more interesting than anything their neighbors had seen. Or all of these motivations and more. In any case, there are round or polygonal barns scattered about the continent without any discernable rhyme or reason.

Probably the greatest of the round barns in North America was also one of the earliest. The Shaker barn at Hancock, Massachusetts, was first built in 1824. After a fire and other problems, it was rebuilt on the original foundation in 1865. It is maintained today by the Hancock Shaker Village Historical Society and looks much as it did when it was rebuilt.

This great stone working monument is 270 feet in circumference with walls that are more than a yard wide. Its roof is virtually flat, topped by a wooden polygon with a window in each of its sides. That in turn is topped by an octagonal cupola in which louvered panels alternate with windows. The cupola crowns an open-sided central structure that acts as a ventilation shaft.

Many smaller circular and polygonal barns are still in use across North America. They are particularly abundant in the American Midwest, with other examples in the Northwest, Vermont, and Canada. Common to most of them is the sense of awe one feels when standing inside. It's a little like standing under the dome of a Renaissance cathedral. Something of the wonder of what man has created in the dome of St. Peter's Cathedral in Rome is replicated in the domed ceiling of a round barn with its necessarily intricate arrangement of rafters and braces.

The composition of the outer walls of these unusual barns runs through almost the entire range of possible barn sidings. There are circular and polygonal barns of stone, brick, and mortared rubble. Others have vertical pine plank siding, others clapboard, and still others are shingled. Apparently, if a person wants to build a round barn, he or she will find a way to do it with the materials at hand.

Most early working farm horses were housed in the barn in the same general area as the cattle. As machinery replaced horse and ox power, however, wealthier farmers kept horses for the pleasure of riding and the specialized stable came into being. Today stables house breeding stock, riding horses, and racehorses. Many have been converted from old barns no longer used as general farm buildings. Most of these, however, have extensively remodeled interiors since horses are social animals and like to see each other. In newer stables, quite different from traditional barns, the stalls are often arranged in a circle or semicircle. Sometimes there are riding rings in the center of the building.

Stables have been built in a great many styles of architecture. They are usually one story with each individual stall having a door that leads directly to the outside. Many of these doors are divided into two sections so that the top section can be opened for the horses to look out. Most stables do have a haymow under the eaves.

ROOFS, DOORS, AND WINDOWS

Since the barn is just about always the largest structure on the farm, and sometimes the largest structure for miles around, it is nearly always impressive in its setting. Most architectural historians acknowledge that much of the awe that barns inspire is due to their size relative to their surroundings, and many have written about the significance of the roof in creating barns' grandeur.

It certainly is the most distinctive feature. The barn roof not only provides shelter by covering the interior space but also defines the silhouette of the barn against the sky. Its shape and material are important factors in distinguishing each barn both aesthetically and functionally.

The first roofing material in North America was almost certainly thatch, a covering made of straw, rushes, reeds, or even leaves. The earliest roof rafters were thin poles, or perhaps roughly shaped tree limbs fastened together at the peak. Other poles or tree limbs were fastened to the original layer to form a crosswise lattice onto which the thatch was piled. Some early immigrants, particularly the Irish, preferred to use sod rather than thatch for their barn roofing.

Because wood was so plentiful in the New World, it soon became the roofing material of choice. Beams (long boards each cut from a single tree) became the rafters and smaller, rough-hewn boards became the horizontal supports. Early wooden roofs had tree bark added to this skeleton. First rows of bark were laid down with the rough, outer side facing the inside of the barn and the concave rows of bark exposed to the sky. On top of this layer, overlapping and connecting the ends of the individual "shingles," rows of bark were laid with their rough, convex surface to the sky. This formed a pattern that looked from a distance not unlike a choppy ocean and that was surprisingly durable and waterproof.

A simpler wooden roof for small barns was built of hollowed-out logs with the bark left in place. Each log would be cut to the exact length of one side of the roof. Then one row would be laid down with the concave side up and another row laid on top of it with the concave side down and the tree bark exposed to the weather.

As settlements became more permanent and farmers had more time, barn roof construction was refined. Pieces of wood were sawed into rectangles with one end thinner than the other to facilitate overlapping. Thus the shingle was born, and became the most widely used roofing material of the nineteenth century. Some farmers who lived near quarries, however, preferred the durability of slate roofs. By the twentieth century, asphalt shingles, aluminum, and steel had all become common barn roofing material.

The gable roof is the simplest and earliest of the barn roof styles. It is formed by two angled surfaces that meet at a ridge line. The word *gable*, however, does not really describe the roof but rather the shape that the roof creates on the vertical ends of the building. A gable is the triangular portion of vertical wall from the eaves or cornice to the ridge of the roof.

All of the earliest New World Dutch barns, English barns, Pennsylvania barns, and southern crib barns had gable roofs. In the lofts under the rafters of these roofs, farmers stored their hay. That hay was stacked tight, filling every available space each fall, and was used all winter until by spring the loft was virtually empty.

The two-story saltbox home with its short roof on one side and its long unbroken roof extending almost to the ground on the other has become a symbol of early American housing, particularly in the Northeast. The same roofline appeared in barn architecture, but not until the nineteenth century and then probably only because the farm was growing and needed more barn space.

Usually an addition was added to an existing barn or extra enclosed space planned for a new one. The space was often needed to house cattle or farm tools or machinery, or for some other purpose on a particular farm. It was added to the length of the barn and then covered by an extension of the roof. This created a barn roof with one short side and one very long side, like the saltbox house from which it took its name.

In houses, the long side of the roof virtually always faces north. This is not always the case in barns, but the longer roof side usually faces the prevailing wind.

The need for more space also stimulated the creation of the gambrel roof. It came into being sometime near the middle of the nineteenth century and rapidly became so popular that it is regarded by most North Americans as the typical barn roof.

The word *gambrel* actually refers to the curved hock of a horse's leg. It seems quite appropriate, therefore, that a barn roof with a bend in its slope should be called a gambrel roof. The somewhat curved shape is caused by the steep lower slope of this roof, while the upper slope was less angled, though not quite flat. This graceful line evolved because as farms got bigger and farmers owned more livestock, they needed more and more space to store enough hay to last through the winter. Since hay was stored under the roof, someone got the idea of raising the roof. A very steep angle from the eaves to a point more or less midway to the ridge greatly increased the hay storage space.

In some barns constructed by immigrants of Dutch heritage, the eave ends of the gambrel roof are curved gently outward, extending the roof beyond the walls of the building. Most authorities believe this was a purely aesthetic variation of the roofline, adding a graceful finishing touch. There is a possibility, however, that this curved addition actually served to carry the rainwater or melting snow that ran down the roof a little farther away from the sides of the barn. By doing this, the water would be less likely to collect against the barn sill and cause moisture-related problems.

With the advent of the use of steel and aluminum in barn roof construction in the twentieth century, the line of the gambrel was refined still further to the rounded form called the rainbow roof, which is seen on many large, modern barns. This construction line has again slightly increased the amount of hay that can be stored under the barn roof.

Hip roofs are probably as old as gable roofs, but they are not often seen in their pure form in barn construction. The hip roof design is also called the cottage roof, after those found in the English countryside. All four sides of the building (which are of equal height) are roofed when hip roof construction is used, so there is no gable. The pitch at the end sections is usually the same as the pitch over the long wall, creating a triangular shape at the narrower ends of the building and a shorter ridge line on the roof.

In barn roof construction, the hip and gable are usually combined in one of two ways. In some barns, the ends start out as gables and then the gable is interrupted with an added triangle of roof. This is sometimes called a snub-nosed roof.

In the crib barns of the south central United States and in some other areas also, barn roofing seems to start out as a hip roof and then suddenly stops on the narrower ends of the barn. A new vertical wall seems to grow out of the barn at about the one-quarter mark on each end, while the roofline continues its angle on the long sides of the barn. The vertical walls that are above the hip roof have a gable. This arrangement is referred to as a gable-on-hip roof.

The design of the mansard roof has been attributed to a seventeenth-century French architect named François Mansard and was probably created out of the same need as the gambrel roof—more space. It is a four-sided roof with a pitch that has two angles. The lower angle, like the gambrel, is quite steep, creating more headroom inside the structure. When the second angle levels to meet at the ridge line, the shape resembles a hip roof. The mansard roof is only rarely seen on barns.

You're probably familiar with the dormer windows set into the roofs of Cape Cod-style houses. They are vertical projections from the roof with a small gable roof of their own set against the pitch of the main roof of the building. In barn construction, dormers sometimes shelter the huge doors to the haymow. Window dormers provide light and ventilation. In barns of the Victorian era, they were sometimes added for purely aesthetic reasons.

In barns of central and western North America, the roof is sometimes extended out several feet over the gable ends to form what appears to be great arrowlike points. These roof extensions were built to shelter the haymow doors.

Since barns were built for strictly utilitarian purposes, it follows that the design of their entranceways should be purely functional and ideally suited to their purpose, which explains the immense variety of barn doors.

Many things had to enter a barn: wagons, animals, people, and the hay and grains that were carried or pitched in. In the evolution of the barn, doors were designed for each function. And most barns have a number and variety of doors.

The biggest doors in the barn had to be big enough to allow a wagon loaded with hay to pass in. In the earliest North American barns, this was accomplished by framing a huge entranceway and then building a separate "door" that was not attached to the entranceway frame. Within this huge panel, a small (relatively speaking), hinged, human-size door would be built.

The wagon door panel would be removed in the spring and the barn left open through the summer and the harvest season. Once the harvest was in, the huge panel would be moved into place and then a beam would be propped against it on the outside to keep it closed and in place. Human comings and goings were accomplished through the door within this panel.

A slightly easier-to-manage version of the huge single-panel wagon door used interlocking metal strips on the top of the panel and the barn to form a primitive sliding door. These strips had to be constantly greased, however, and even when greased, the door was often difficult to move. As a result, it was generally left closed. Within this huge panel, there was usually at least one other door (human size) and sometimes a second door (farm animal size).

In the 1840's a design for freight car doors on rollers was grabbed up by farmers. Then the huge panel wagon door could be moved to the side more easily. Today roller doors are among the most commonly used large entranceway doors.

Most Dutch and English style barns covered the wagon entranceway with two door panels on hinges that allowed each door to swing open 180 degrees and rest against the barn walls. Because each of these doors was large, they had to be built for strength and stability and to avoid warping. The traditional X or Z bracing now considered a decorative characteristic of barn doors was originally used to keep them straight and steady.

Haymow doors were usually located just under the peak of the gable although some barns, particularly the Quebec connected barn, actually had dormers built into the roof for the haymow doors. These were not as large as wagon doors but big enough to allow plenty of working space. In early barns, as today, they were usually double doors, hinged to open flat against the barn. Some haymow doors, however, are built in one huge section and hinged at the bottom so that when opened they drop flat against the barn below the haymow opening. A few are roller doors just like the main wagon doors of the barn.

In Pennsylvania barns where the animals are housed on a separate level, there are often several doors to allow entry for individual animals, similar to modern stables. Many of these were what is now called the Dutch door, a door split horizontally allowing the top half to be opened while keeping the lower section closed. This served the purpose of allowing in light and ventilation without letting the animals out of the barn.

Even in the earliest New World Dutch barns of New York State, there are often human-size doors on either side of the wagon doors. In the connected barns of Quebec, there is a separate door to each section of the barn appropriate to the use of that section. And some sections of the connected farm buildings of New England actually have no external door.

Many early barn doors were covered by a hood or small roof. This was probably done as a shelter against inclement weather for the farmer and also as a means of keeping the door sill as dry as possible to prevent its rotting.

Several millennia from now, archaeologists on a dig might come across the remains of yet another kind of door in a North American stone barn and wonder. Why would a farmer build a very small (perhaps eight inches high and six inches wide) framed opening in the stone foundation? They might get a clue if nature has been extraordinarily kind and preserved the hinges and perhaps part of the wooden door that could swing inward or outward. Perhaps one of the archaeologists will realize that this was the doorway of some very important farm animals, the barn cats. No such door will be found in old wooden barns where the construction usually left many places for wily mousers to squeeze through.

Ventilation in very early farm buildings was left to nature. Often light (and air) passed through the shingles of the roof which would swell in rainy weather to form more waterproof covering. Early vertical barn siding was rarely tongue and groove and left plenty of space for air to flow. So did early clapboards and, of course, rough-hewn logs. As a result, barn windows were rare.

Early New England farmers, however, did ensure ventilation and light with long, narrow transom openings above the wagon doors, running across their entire width. These openings were covered by a small hinged flap that could be opened or closed according to the weather and the farmer's need for light. Later, when glass became readily available and inexpensive, the rooflike cover over the door ventilation slots was replaced with a row of glass panels. Many of these New England door lights can still be seen on barns in the area.

Ventilation for the haymow was often created by carving pigeon holes at the peak of the gable ends. Pigeons were abundant on early eastern farms and their presence was encouraged since they often substituted for chicken on the farmer's dining table. Other birds were also encouraged to nest in the barn. In nineteenth-century barns that are still standing, you might see as many as six so-called "marten holes" just under the peak of each gable. They were almost always arranged in a triangular pattern. When there were more than six holes, they might be a random arrangement of different sizes as though inviting a variety of birds to use the barn.

In Pennsylvania barns, the stone sides were often built with long narrow slits called loophole windows. Smaller round holes in the stone, usually lined with bricks, were also called loops. Both the slits and the round holes were flayed outward on the inside of the building. This construction was extremely effective because it let the warm air out while preventing rain from getting in.

In barns built of brick, many builders let their creative spirits explode. They often provided ventilation by leaving out selected bricks. The selection of missing bricks was not random, however, and the patterns created by the empty spaces are often aesthetically pleasing and/or symbolic. In some cases, they look like quilt patterns.

Early barn windows were not what we know today. They were simply open spaces cut into the wall that could be covered by removable panels in bad weather. These holes were placed with nothing but function and need in mind and form no discernable pattern on the exterior of the barn. As early glass windows were added to barns they also followed no predictable pattern or shape.

Even today, few barn windows are double hung to be raised and lowered like house windows. Some simply provide light and cannot be opened. Others are hinged on the top and can be propped open, while others slide from side to side. Sometimes barn windows serve no other purpose than an ornamentation, as in the occasional octagonal or circular window set high in the gable of the barn like the rose window of a church.

Some barns, particularly in more southern regions of North America, have louvers rather than windows. These provide ventilation with protection from rain. Inside the barn, however, only diffuse light comes from the louvered openings.

The "curing" of green hay can generate enough heat to start a fire through spontaneous combustion. As farmers learned this, they began to search for ways to improve haymow ventilation so that the warm air could easily escape. Since pigeon and marten holes were not enough, ventilation shafts were added to some barns and planned in the construction of others. These were often topped with cupolas.

CUPOLAS

Cupola comes into the English language from the Latin *cupola* meaning "small tub," which seems to have but little relationship to the ornamental structures that top some barn roofs in North America. In later usage, however, cupola came to mean "small vault," and the word was used to refer to smaller, secondary domes in a cathedral, usually built to add light to an area. The structure that became the North American cupola was hardly a dome, but it was a topping for the most important building on the farm and it did sometimes add light to the area beneath the rafters.

Barns in Connecticut were probably the first to add cupolas. The earliest ones were strictly for ventilation. Usually they were four-sided structures with small gable roofs and louvered openings on each side. The ventilation worked and the idea stuck and spread throughout New England. But cupolas changed from merely functional to a combination of functional and decorative as farmers went to work using this barn topping as a statement of personal craftsmanship and artistic expression.

Some cupolas were built with gables on all four sides to form a starlike pattern, others were built with slender curved roofs rising like church spires, while still others were actually topped by domes. Sometimes copper sheathing was used as the roofing material, although shingles, slate, or even tiles were more common. The shapes of the cupolas themselves vary tremendously. Some are square, some rectangular, some hexagonal or octagonal, and a few are circular.

Depending upon the taste of the farmer, the cupola might have classic mouldings, latticework, or a Victorian gingerbread of posts, poles, curved cornices, and open cutwork. Later cupolas sometimes alternated double-hung windows with louvers and some included marten holes for small birds. Some wealthy farmers even installed clocks in these little structures atop their barns.

ADORNMENTS

People have adorned their homes and workplaces since

the time of the cave dwellers, and their love of adornment didn't stop with the cupola on top of the barn roof. Many farmers added weather vanes to the cupola. These wonders of American ironwork range from simple arrows to point the way the wind blows, to designs ranging from the sublime to the downright funny.

Horses are by far the most popular weather vane subjects but roosters, bulls and cows, dogs, deer, and even beavers, fish, and whales have been done. The base of the weather vane might be a simple pole or an intricate scrollwork with arrows and the initials for the four points of the compass.

But barn decoration didn't stop at the top of the barn, either. In Bavaria and Austria, murals are routinely painted on the sides of houses and barns. American farmers were not so likely to paint their family crests or biblical scenes, like Moses receiving the Ten Commandments, on the buildings they owned, but they did paint their farm animals. Again, horses are the most popular image, usually painted in side view in the American primitive style.

Almost as often as horses, you'll come upon barns with cattle painted on the barn walls that are adjacent to the wagon doors. Or sometimes a great bull will be painted under the peak of the gable. It's as though the proud farmer is showcasing his best animals.

In Quebec and parts of Ontario, many farmers have painted their barn doors in bright colors and original patterns. Geometric designs, flowers, clover, wheat, traditional quilt patterns—all can be found on the barn doors in the more northern reaches of North America.

Farther south in the central United States, and in Pennsylvania in particular, many barns are decorated with colorful painted circles called hex signs—six pointed stars, swirling crosses, flowers, hearts, or playful geometric designs in a wealth of colors and sizes. In the Middle Atlantic states, they have become such a part of the farm heritage, that visitors can buy mass-produced copies in many antique shops and most tourist traps.

But the original ones were probably made by German and Swiss immigrants who were Mennonites or Amish. The word *hex* comes from the German *Hexe,* meaning witch. In current American usage it usually connotes a spell being cast—a kind of bewitching. Some authorities believe that the colorful hex signs were painted on barns to protect the cattle from evil spirits. Today, however, most Mennonite and Amish farm folk will tell you the paintings are "chust for pretty."

The six-pointed star and the swirling cross or swastika, however, can be traced back to ancient Indo-European cultures, usually as symbols of good will and well-being. Painted on the sides of the largest buildings on a farm, they may be a welcoming and protecting symbol. They certainly are a tribute to human love of color, line, and beauty.

Red is the favorite color for painted barns in North America, with white close behind. Why are so many barns red? Perhaps because it was also a favored color in Europe, but more likely because it was the easiest color to mix on the farm. Into the early nineteenth century, paint was difficult to obtain and most farmers made their own. The most popular mixture was a blend of readily available red iron oxide and skim milk (often left over from the churning of butter) with some lime and linseed oil (or cow's hoof glue) added. There are some stories that farm animal blood was also mixed into the paint to deepen the color, but this is doubtful.

In early southern settlements, lamp black was used to create paint that seeped into the wood like our modern wood stain. The resulting colors were usually subtle shades of gray. More popular, however, was the use of ocher (called *oaker* at the time), an earthy clay. When mixed with hematite, the color tends toward red; when limonite is added, toward yellow.

Early wooden barns in the northern areas of the continent were rarely painted, but instead were allowed to weather naturally. Those that are still standing have a remarkably varied and beautiful gray patina sometimes toned with honeylike hues or shaded with dark umber. Dark-stained barns—black or brown—are also popular in some areas, particularly around the Appalachian Mountains.

Many barns have carefully painted white trim around the doors and windows and sometimes on the bracing beams of the doors. This aesthetically pleasing detail had its origin in religious superstition. In his book *The New World Dutch Barn,* John Fitchen quotes from a Pennsylvania folklore publication, "To keep the devil out of your barn, paint a white line around your barn door much higher than the door itself. When the devil opens the door, he will not stop and will run against the line. He will not return." There's hardly a farmer alive today who will attempt to ward off the devil with white paint, but the handed-down tradition of painted lines has certainly added beauty and interest to the landscape.

In the early years of the twentieth century, the sides of barns were sometimes used for advertising. Company representatives would offer to paint a farmer's barn if the company message could be painted on the side facing the road. Some of these advertising paintings still survive, and the modern-day traveler of back roads can read about the best brands of chewing tobacco or horse liniment.

Some barns, particularly in Quebec, include special places for religious grottoes. Statues of Christ, Mary, or a special saint are usually framed in the hewn beams and covered with a small roof. Some of these are located inside the barn, while others are set into the roof in dormerlike fashion. They stand as a simple and beautiful statement of a farm family's prayers for their stock and the harvest.

Late in the second half of the nineteenth century, the intricate woodwork trim of Victorian architecture found its way to the barn. Arches, peaks, fretwork panels, and gingerbread wood carving were added. The effect is a delicate, almost feminine aspect. Such Victorian barns may be every bit as efficient and functional as any other, but they also make a statement of pride and beauty.

SILOS

Say "Think of a barn" to Americans or Canadians and many will imagine not only a commodious building but also a round tower standing nearby. The silo is much a part of our concept of "barn." However, vertical silos didn't come into general use until the very end of the nineteenth century. Illinois claims to be the home of the very first wooden upright silo, built in 1873, and Michigan claims the first stone silo was built there in 1875. Early wooden silos had problems with cleanliness, mold, fires, and windstorms, and most are no longer standing. Stone was more durable, but while there are many early stone silos in Wisconsin as well as in Michigan, the dates of their construction are guesses at best. What we do know is that by the turn of the century, the concept of storing green fodder in an airtight tower had spread across the continent.

The word *silo* probably comes from the French *ensiler*. Early French silos were pits or trenches into which green crops were pressed and sealed off while they fermented. In North America, early silos were also pits protected from weather by thatch or boards, but they were used as storage areas for corn and potatoes or other root crops.

When the idea of storing animal fodder began to take hold, the earliest stone silos were as deep into the ground as they were tall, which meant they were squat little structures. Then someone realized that a stone structure with thirty-inch-thick walls was just as airtight above ground as below, and had little chance of toppling over no matter how high it was. The digging of deep pits was abandoned and building practices that led to our modern stone, brick, and concrete structures were developed.

The earliest wooden silos were slats held together with wire loops, constructed much the same way barrels were once made. These towers were extremely unstable, however, and tilted to a different angle with each windstorm. They also were rarely airtight, which led to much spoilage. It is extremely unlikely that any survive today. Other experimental silo styles were built *inside* the barn, sometimes as square bins in the corner and sometimes as the central core of a round barn. Outdoor towers, however, proved more efficient and cost-effective.

Most modern silos have domed roofs, but these are the product of modern construction materials and techniques. Early roofs were cones or polygons of wood shingles or slate. Some had a dormer or two, some were topped by cupolas, and some emphasized their peaks with lightning rods or weather vanes.

INSIDE THE BARN

In most of the structures that people have created, the interiors have been carefully smoothed. Space is divided into rooms and, for the most part, the working systems and structural elements are covered by similar or at least harmonious surfacing materials. Only inside two kinds of buildings can one observe, study, and feel how the building was made. These two buildings are the Gothic cathedral, where every stone helps to create a structure that will soar to the greater glory of God, and the barn, where every beam and post helps create a structure that will house and protect the farm products and animals that provide sustenance for humanity.

Various elements of the two architectural forms have been compared: their floor plans, their vastness, their grandeur, the play of light, the reverberation of sound. But it is in the breathless awe that an observer feels when seeing the ordered space created by human hands that these two buildings, so dissimilar in function, are most similar.

In the barn, the central aisle, or bay, is usually a threshing floor. Like the central floor of the cathedral, it is worn with time and footsteps, but there is yet another dimension in the barn. The floor is worn with the work of the farm, often worn to a smoothness and subtle coloring that would rival marble.

In the age when threshing of wheat, oats, rye, and barley was done by manual labor, the barn was the factory of the farm. After being harvested, the sheaves of grain were laid on the floor of the wide central aisle. They were threshed by the farmer, his family, and his hired hands with a flail, a tool that ended in an attached wooden arm. The idea was to crush the kernels and split them off from the stalks which would become straw for the animal's beds. The grain was then separated and stored for the next step, winnowing.

Winnowing had to be done on a windy day. Doors at both ends of the barn were opened and the wind was invited to whistle through. The farmers used this natural force to separate the crushed kernels from the husks or chaff. The threshed grain would be loaded into a wide but shallow tray and tossed into the air. Since the husks and chaff were lighter than the grain, they flew up and blew away and the grain dropped back into the tray. As each tray load became free of its undesirable elements, the grain was poured into its storage bin and a new batch of threshed grain was piled up for winnowing.

This process of threshing and winnowing was repeated every year and each year the threshing floorboards took on a softer sheen. At other times on those same boards dancing feet replaced pounding flails. And sometimes children played there and women sometimes spun flax or wool. In the boards of the threshing floor of every barn is the history of the farm and the family, its sweat, tears, blood, and the echo of music and laughter.

In the narrower aisles on either side of the threshing floor, animals were housed or grain was stored. In fact, the width of a barn bay has been determined for more than two millennia by the space needed to house oxen. Since Roman times, the space allotted to house oxen has been seven feet for a pair or four feet for one. The bays of a barn are usually built to house two oxen. The bents of most barns are seven to eight feet apart and that is therefore the width of each bay. If animals were not sheltered in the barn, the space in the narrower side aisles was usually allotted to grain bins, but the size of the bays remained about the same.

Hay was also stored in the loft above the threshing floor. Poles were laid across the great anchor beams that ran from post to post and the hay was pitched up in the fall. It was an easy matter to drop it down to feed the farm animals.

In the ordered space of the barn, one is surrounded by the beauty of wood. The marks of an ax or adz that was

swung a hundred years ago still give a painterly texture to the beams and posts. A thousand shades of brown, gray, and other earth tones rise up in the wood in response to the ever-changing light of the barn.

An oak peg, driven into a post seventy years ago to hold a harness, may still be intact. There are no nails in the oldest barns, only carefully fitted pieces held together with oak treenails. The farmer trusted the fine wood. Even the ladders were often nothing more than large oak pegs set into log siding. Hooks for hanging harnesses or coats were of the simplest construction: the crotch of a tree where a small branch separates from a larger branch.

The barn is a place of sight, sound, and smell. Wisps of hay, the warmth of animals, the play of light, the sounds and occasional flights of birds. Above all it is a place, or perhaps a palace, of life.

*Opposite: This small English barn is patterned after early eighteenth-century New England construction. It has a Victorian cupola and a tiny window under the peak of the gable. **Following pages:** Some American farmers took the traditional English barn plan and let their creativity rule. This playful structure has candy-striped board and batten siding and a flared Dutch gambrel roof. There's no reason except artistic accent for the gabled dormer over the wagon doors, which are asymmetrically balanced by a mid-size door, a small door, and scattered windows.*

English Barns

This unpainted Yankee barn could be two hundred years old. The windows, however, were almost certainly added in the twentieth century. **Below:** The wagon door in the center of the long wall betrays the English heritage of this barn, but the farmer has expanded the traditional plan with a lean-to at one end, some extra doorways, and a cupola.

What was once the opening for the wagon doors on this Dutch gambrel barn has been converted to something like a garage door with smaller doors on each side. **Overleaf:** *This traditional English barn, with a pent roof over its wagon doors, has two lean-to additions and a rare silo built of wooden slats held together by wire rings.*

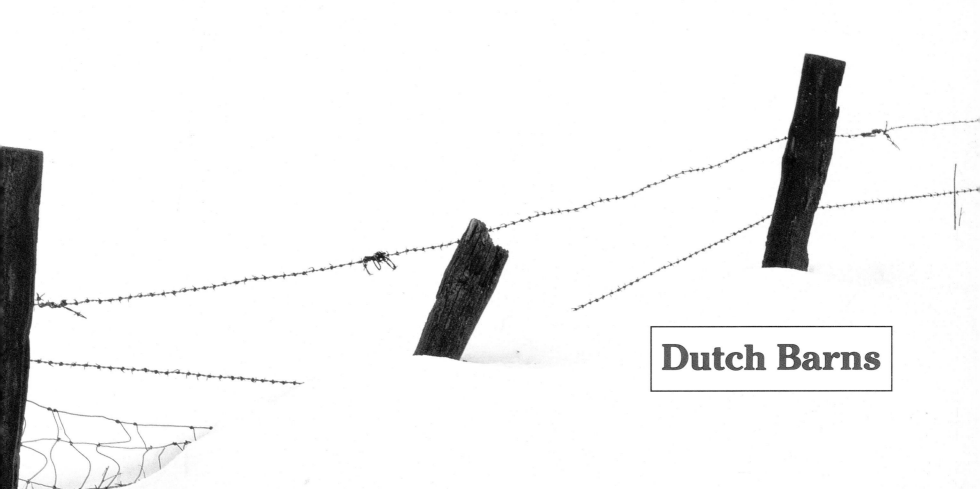

Dutch Barns

Preceding pages: *The earliest barns built on the European continent were said to resemble the hulls of overturned ships. Standing in the drifting snow, this Oregon adaptation of ancient barn architecture, with its weathered boards blending with the sea of gray and white that surrounds it, brings to mind the ship image.* **This page:** *Long after Dutch barns were no longer being built in the East, this pioneer adaptation of the style was created in Methow Valley, Washington.* **Opposite:** *Rows of windows, a silo, and ventilation hoods all characterize the American dairy barn.* **Following pages:** *Lean-to additions have been built on both sides of this basically Dutch barn, windows have been cut in at random spots, and there is a door to fit seemingly every demand.*

Preceding pages: In this old Pennsylvania bank barn, farm animals were housed on the ground level, protected from both the winter wind and the summer sun by the stone construction that was set into the earth for the length of one wall. **This page:** *This stone barn in Bucks County, Pennsylvania, is actually an English barn, but earth has been built up to create a ramp to the wagon doors on the second floor.* **Below:** *The man-made ramp leads to the second-floor threshing area of this English barn.* **Opposite:** *Most bank barn foundations were made from stones found in nearby fields. Wooden doors with forged hinges were set into the stone.*

Preceding pages: *Constructed from local fieldstone, the wall of this barn has a mosaic quality.* **This page:** *This Pennsylvania bank barn, built in 1875 of limestone blocks, resembles a small European castle more than the early wooden barns of the U.S.*

Many brick barns in Pennsylvania and the Midwest were ventilated by leaving out selected bricks during construction. Few masons omitted bricks randomly, however, and beautiful decorative patterns emerged from the ventilation work.

Preceding pages: One section of this bank barn in New York is built perpendicular to the other, forming a sheltered barnyard. **This page:** This 1896 barn in Wood County, Ohio, is a proud Victorian with peaks, gables, pigeonholes, louvers, and a lightning rod. **Opposite:** Another Victorian, this 1897 barn in Garrett County, Maryland is as eclectic as its era. It includes two gable roofs, four cupolas with four gable peaks each, and elements of bank and English barns.

German settlers in Somerset County, Pennsylvania, created the fretwork decoration on this barn, which dates from 1880. **Opposite:** *Delicate trim edges the roofline of this barn in the Amish country of Pennsylvania. Louvers in the walls and the cupola serve as ventilators.*

Two gabled dormers, each with a tiny window, and two red cupolas protrude from the black gambrel roof of this unusual bank barn.

Built in 1900, this Portage County, Ohio, barn is trimmed in white, which was once thought to keep evil spirits out of barns.

Purple martin holes under the gambrel roof peak invite nesters to this bank barn in Polk County, Pennsylvania.

*Posts of solid walnut support the cantilever in the Pollins Barn. Built in 1849 in Westmoreland County, Pennsylvania, the building is an example of early bank barn construction. **Below:** Unpainted horizontal and vertical siding gives this bank barn a mottled look. The cantilever beams are visible above the stone foundation. **Overleaf:** Twin silos balance the twin ventilation hoods on the gambrel roof of this barn.*

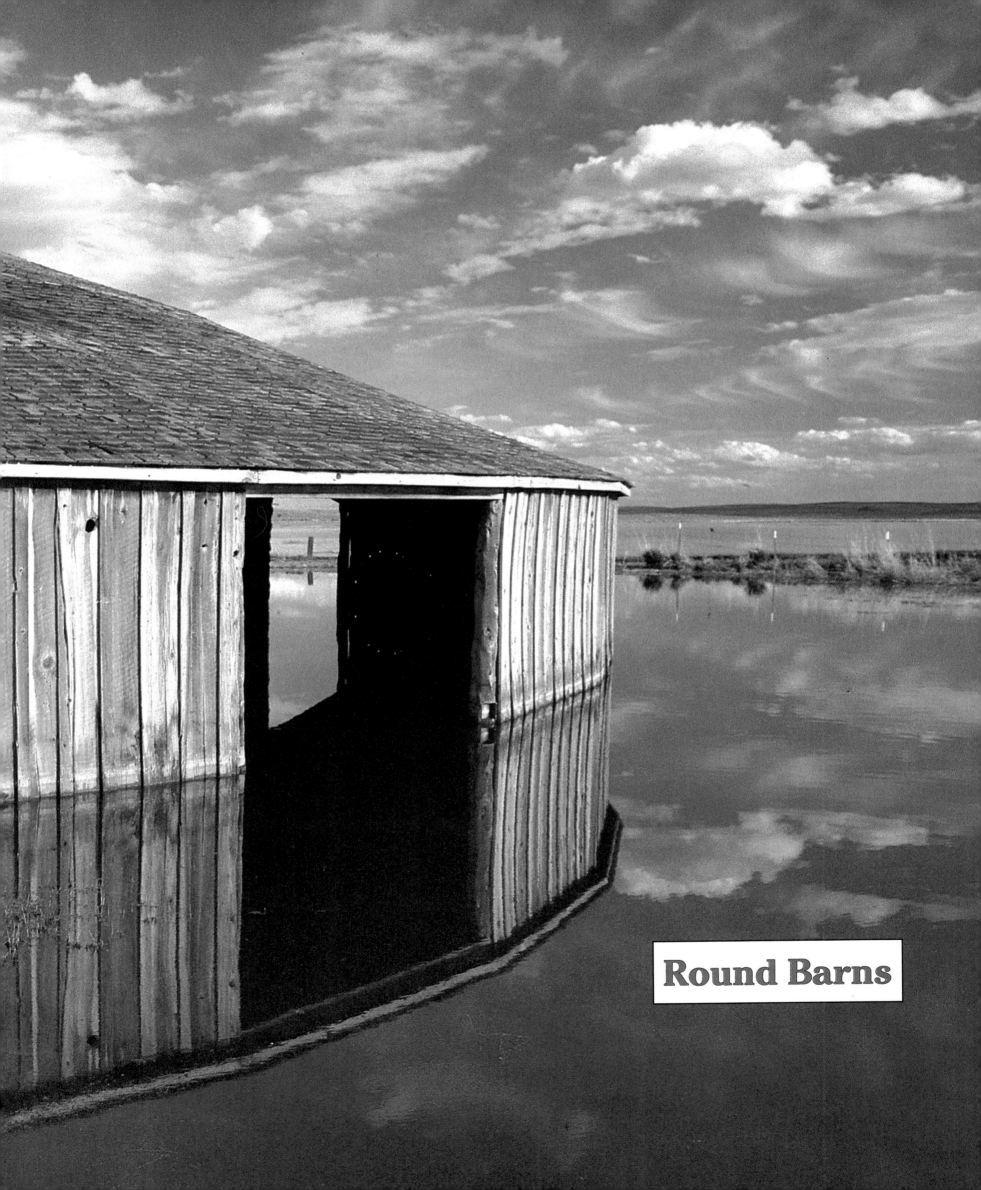

Round Barns

Preceding pages: *In search of the perfect form, some farmers built round barns like this one in Oregon.* **This page:** *The roof rafters go around and around this barn in Marion County, West Virginia.* **Below:** *Supports brace the roof of a round barn on Bainbridge Island in Washington.* **Opposite:** *The play of light on the rafters creates a dizzying effect.*

The Shaker stone barn in Hancock, Massachusetts is the oldest round barn in the U.S. First built in 1824, it was destroyed by fire and then rebuilt in 1865 on the same foundation. **Opposite:** The barn, 270 feet in circumference with walls over a yard thick, is a part of the Hancock Shaker Village, which is open to the public. **Following pages, left:** This 11-sided wooden barn has a gracefully curved gable roof over the dormer, which shelters the haymow door. **Right:** The gently flared roof on this three-story octagonal barn is repeated in miniature on the cupola.

Preceding pages: This focal point of the farm dwarfs the young trees that are its near neighbors. **Opposite:** This Vermont round barn uses the bank barn principle and shelters its entranceway with a structure not unlike a covered bridge. **This page:** In the early years of the twentieth century, round barns like this 1921 model became very popular in the Midwest. **Below:** Brick, wood, glass, and shingles have been combined to create a sturdy, timeless, monumental barn on a farm in Ida Grove, Iowa.

Connected
Barns

Previous pages: *The weather vane, with its cow motif, overlooks real cows in the barnyard.* **Preceding page:** *There's an asymmetrical beauty in the intersecting angles of these connected New England farm buildings.* **This page:** *Since no farmer wanted to trudge through the snow of a northeastern winter to feed the livestock, connecting the farm buildings seemed a logical answer to the problem. These barns are in Woodstock, Massachusetts.*

*Built in the Quebec connected barn style, each of these long, low barns is divided into rooms serving different functions. **Opposite:** Sometimes the connected farm buildings of northern New England form great zigzags along the roadside.*

Log Barns

Previous pages: *Crib barns like this one were used for storing grain.* **Preceding pages, left:** *Some farmers stuffed the spaces between the horizontal beams of log barns with mud or grass, but the builder of this one chose to leave the wood unchinked.* **Right:** *This crib barn with a gable roof was built of logs in 1895.*

Preceding page, top: This Green County, Tennessee, crib barn has been in use since 1895. Expansion was simply a matter of adding lean-tos. **Bottom:** *Landis Valley, Pennsylvania, is the location of this reconstruction of a 1790 English barn. The log walls are chinked with natural clay, and the roof is thatched.* **This page:** *Abraham Lincoln's farm in Coles County, Illinois, is the home of this double crib barn.*

Top Hat
&
Crib Barns

Preceding pages: *This type of broken gable roofline gets the nickname "top-hat."*
Opposite: *Northeast of Arthur, Iowa, this once-elegant top-hat with an added tower has given in to wind and weather.* **Above:** *The subtle, time-rubbed colors of these boards cannot be duplicated with wood stain or paint.*

This lonely pair stand on the Arthur Ranch in Indiana.

*This lively top-hat barn has candy-cane sides and a mural under the rain hood. **Right:** Many unpainted barns with weathered metal roofs, like this one at Cades Cove Historic Park in Tennessee, still dot the landscape around the Great Smoky Mountains. **Overleaf:** The double crib barn with cantilevered haymows and a drive-through wagonway is found almost exclusively in the Southeast.*

Western Barns

Preceding pages: *Evening winter light plays on this barn in Union County, Oregon.* **This page:** *A tiny stone barn in the American Northwest echoes the mountain peak in the distance.* **Below:** *This buckling horse barn looks as though it belongs among the dilapidated buildings of a western ghost town.*

The colors, patterns, and placement of the logs that were used to build this rugged barn tell the story of its growth at the Mormon Row Historic Site in Jackson, Wyoming. **Below:** These three log barns stand in Story, Wyoming.

Preceding pages, left: *Many western barns started as single-story log buildings and then grew upward as the farmers prospered.* **Right:** *Log construction is a good insulator against both hot and cold weather for the animals housed in the ground floor section of this western barn.* **Above:** *The owner of this barn near San Luis Obispo, California, may have had his roof paid for by the Mail Pouch Tobacco Company. Many farmers along America's highways bartered barnside advertising space for a new paint job, siding, a lean-to, or roofing.*

It's salt air that weathers the boards of this barn near Gull Rock at Sonoma Coast State Beach in California, but it also creates the same soft silver hues that characterize the lobster huts of the Atlantic Coast.

Horse Barns

Preceding page: *A Kentucky horse barn, painted as boldly as a jockey's shirt, stands proudly beneath the shade of great trees.* **This page:** *This Amish barn plays host to the horses and carriages of churchgoers in Lancaster County, Pennsylvania.* **Below:** *Louvers stand in for windows in this York County, Pennsylvania, barn, built in 1890.* **Opposite:** *A horse barn in Palo Alto, California, has miniature gables on the cupola, as well as unusual corner windows.*

Built in 1955, this Pennsylvania barn seems to have collected features from different times and regions, but the louvered cupolas over the ventilation shafts are definitely a twentieth-century touch. **Below:** A 1940 horse barn is topped by a gambrel roof with a long, almost flat extension.

The Dutch doors of this barn are divided horizontally. Top sections are opened to allow air and light into the barn while keeping the animals safely inside. **Below:** Stables often have Dutch doors on each stall. Being social animals, horses like to have the tops of the doors left open so they can look out to see each other.

Tobacco Barns

Preceding pages: Ventilation is the most important factor in tobacco barns, which often have wooden door latches that hold the hinged ventilating doors closed. **This page:** *An Amish tobacco barn stands in Lancaster County, Pennsylvania.* **Opposite:** *Individual hinged panels of siding swing out to provide proper ventilation for the drying leaves.*

Few people think of Ohio as a tobacco-growing state, but this barn in Brown County looks as though it's filled to capacity. **Below:** This barn has been used for drying Kentucky tobacco since 1910. Wagons loaded with freshly picked tobacco leaves can be driven in through the gable end door, unloaded, and driven out through the door at the opposite end. **Opposite:** This Pennsyvlania Amish barn is used for drying tobacco and storing corn.

Roof Types

Preceding pages: This Bernard, Iowa, barn has an open haymow door under the peak of its gambrel roof. **This page:** The four gables of this square barn form a starlike pattern, but each is a little different from the others. One is a simple gable, one has a pointed rainhood, and the one above the wagon doors has a roof extension that shades the front of the barn. **Opposite:** Twin gable peaks top this big, red barn, and they in turn are topped by a dome that is topped by a square cupola with a cottage roof.

Hipped gable roofs are not often seen on barns, but this one works well in California, and provides shade as well as shelter. **Below:** As barns were expanded, rooflines were extended. The main part of this barn has a gable roof that is extended over the addition.

This Missouri barn once was a graceful presence on the land, but today only its skeleton survives. **Below:** This saltbox barn is unusual because the short side of the roof is broken by three gabled dormers and the ridgeline is topped with two ventilation hoods. **Overleaf:** A Michigan barn has a snub-nosed gable roof made entirely of tile.

*Preceding page: Pointed rain hoods like the one on this Montpelier, Virginia, barn first became popular in central and western America and then moved back across the Applachian Mountains. **This page:** Springtime fringes the straight, unadorned lines of this well-maintained barn in the American Northwest. The eaves near the peak of the gambrel roof are just visible. **Below:** Missing roofing material on this barn in Steamboat, Colorado, exposes the roof skeleton to the sky.*

Preceding page: *Pointed arches usually associated with cathedral construction are repeated here in the windows, the roofline, the dormer, and even the cupola. White stucco walls make it even harder to believe this building is really a barn.* **This page:** *If heraldic flags were flying from the tops of the silos and cupolas of this Leeland County, Michigan, barn, people just might mistake it for a castle. Curved dormers with eyelike windows look out from the tops of the silos on either side of the wagon entrance ramp.*

Preceding page: *Soft gray board and batten siding helps to establish the quiet, simple dignity of this Mennonite barn in Goessel, Kansas.* **This page:** *The livestock door on this Kearney, Nebraska, barn moves from side to side on a roller track. The haymow door under the rain hood is hinged at the bottom to drop down out of the way while the mow is stocked.* **Below:** *Snow tops the silos and the barn on this Colorado farm.*

Details
&
Decorations

Preceding pages: *The wood siding of this barn in Ontario, Canada, has faded to a variety of warm colors.* **This page:** *A metal roof tops this Arkansas fieldstone barn.* **Below:** *The marks of an adze can still be seen in the hand-hewn lintel above the window.* **Opposite:** *Sunlight pulls the rich brown tones out of the siding of this Sussex County, New Jersey, barn.*

*Preceding pages, left: A Virginia top-hat barn contrasts with the autumn foliage. **Right:** Age has made the haymow doors of this barn in West Fort Ann, New York, more beautiful, accenting the interplay of color and line. **Opposite:** Indelible stains mark changes long past in the wall of this Walts River, Vermont, barn. **This page:** Spanish influence is evident in this old stone barn at the Lyndon B. Johnson National Historic Park in Blanco County, Texas. **Below:** Stone, plank, and unchinked log construction are also on display at the Johnson Settlement.*

A wooden door fastener is augmented by a more modern latch. **Below:** Some very old, hand-made wooden door fasteners are still in use today. **Opposite:** Huge barn doors had to be carefully braced to keep them straight and true. This door opens to a barn converted to a firehouse in Milan, New York.

 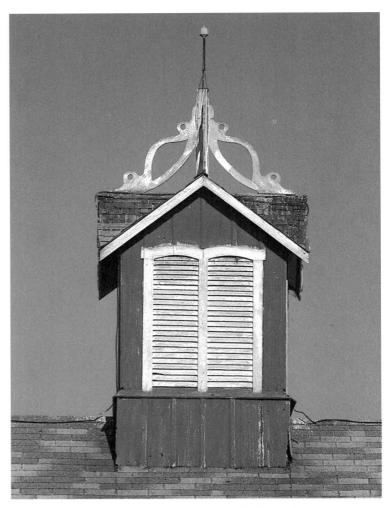

Preceding pages, left: The Stillwater Trading Post in the state of Washington was once a working barn. Windows like this tip window above the sign were often set under the peak of the gable in early New England barns. **Right:** Even the cupola of this barn has four gables. **This page, above:** Louvers on the cupola decorate and ventilate this barn. A haloed window under the peak of the main gable adds light to the loft area. **Right:** There is an artistic spirit in some men that will send them climbing to the cupola at the top of the barn to make a few graceful lines against the sky. **Below:** On a hilltop near Stowe, Vermont, these farm animals feel at home near their cedar-shingled barn.

This page, above: The horse is the most popular weather vane subject in North America. This one, hand crafted in copper, stands atop a barn in Missouri. **Right:** A larger-than-life ear of corn decorates the top of a silo near Perry, Iowa. **Below:** The weather vanes that top the ventilation hoods of this dairy barn sport cows.

This page: *In Bavaria, many people paint murals on the sides of houses and farm buildings, but this American farmer's painting could compete with any Bavarian's.* **Opposite:** *Farming may butter his bread, but this farmer has something to say about the place of music and art in his life.*

Preceding page: Although these symbols on a Pennsylvania Dutch bank barn are called "hex signs," they're really just for decoration. ***Above:*** Intricate German fretwork is characteristically seen on barns in several regions of Pennsylvania.

Built in 1890, this barn in Blair City, Pennsylvania, has unusual decorative motifs. **Below:** This Pennsylvania bank barn, built in 1920, has been recently redecorated. Green louvers have been added to the traditional red-with-white-trim color scheme, and scalloped woodwork ornaments the gable ends of the roof.

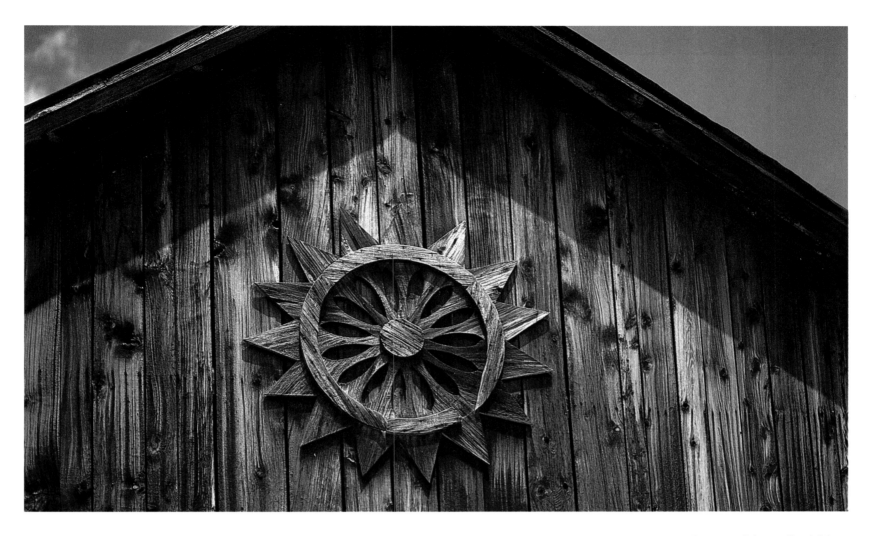

Naturally weathered wood in an intricate sunburst design decorates this gable peak. **Below:** The original louvers are still a part of the walls of this 1894 barn. **Following pages, left:** There's no question about the age of this cedar and red oak barn. **Right:** These white barn ladders stand out in the gloom of the barn.

__Preceding pages:__ Rafters, posts, and beams frame bits of the sky.
__This page, above:__ A barn fire is a sight of terrifying power and a
scene of overwhelming loss. __Opposite:__ "In me thou see'st the twilight
of such day / As after sunset fadeth in the west . . ."—William
Shakespeare, Sonnet 73.

INDEX OF PHOTOGRAPHY

TIB indicates The Image Bank